KING SNAKE

UNDERSTANDING AND CARING FOR KING SNAKES AS PETS

DR MORRIS HART

Copyright© 2024 **DR MORRIS HART**

All rights reserved. No part or part of this book or publication may be reproduced, stored, or transferred in any form by electronic, mechanical, recording, or other retrieval system without written permission from the publisher

Table of Contents

INTRODUCTION **5**

CHAPTER 1 .... **15**

SELECTING THE IDEAL KING SNAKE: TYPES AND FEATURES 15

CHAPTER 2 .... **23**

SETTING UP THE PERFECT ENVIRONMENT AND HOUSING 23

CHAPTER 3 .... **35**

FEEDING YOUR KING SNAKE: NUTRITIONAL AND DIETARY ADVICE
.... 35

CHAPTER 4 .... **46**

TAKING CARE WHEN HANDLING KINGSNAKES: SAFE HANDLING PRACTICES 46

CHAPTER 5 .... **56**

HEALTH AND WELL-BEING: TYPICAL PROBLEMS AND APPROPRIATE TREATMENT FOR KING SNAKES 56

CHAPTER 5 .... **66**

Sophisticated Maintenance Advice for Skilled King Snake Keepers 66

CHAPTER 6 ....**79**

Reproduction and Breeding Considerations in the Breeding of King Snakes 79

CHAPTER 7 ....**96**

FAQs for King Snake: Common Questions Addressed 96

CHAPTER 8 ....**111**

In conclusion 111

Introduction

Reptile fans have been enthralled with king snakes for generations due to their magnificent appearance and eye-catching hues. Many pet owners have a particular place in their hearts for these snakes because of their beautiful appearance and intriguing activities. We will delve into the world of king snakes in this extensive book, examining their natural history, habitat, behavior, and maintenance needs. Gaining knowledge about these majestic reptiles will let you provide your scaly friend the finest care possible.

- The Evolution of King Snakes

Knowing the natural history of king snakes is crucial to appreciating them to their fullest. These snakes are members of the Lampropeltis genus, which is distributed across North and Central America and has a number of species. The common king snake, Lampropeltis getula, is one of the most iconic species.

King snakes are well known for their remarkable range of adaptations and extensive range. They live in a range of environments, such as urban areas, meadows, woodlands, and deserts. Their distribution reaches as far south as Central America from the southeast region of the United States.

The food of king snakes is one of their most interesting characteristics. They are strong constrictors that mostly eat other reptiles, such as snakes, lizards, and even poisonous ones like rattlesnakes. Given that they are regarded as the kings of reptile hunting, this food predilection has earned them the moniker "king" snake.

- Characteristics and Morphology

King snakes are well-known for their eye-catching patterns and colors. Although there is a great deal of variety across species and subspecies, they usually have bright stripes or bands that stand out against a darker background. In their natural habitat, their unique patterning acts as camouflage, enabling them to blend in with their surroundings while seeking prey.

King snakes are distinguished by their vivid coloring as well as their sleek, muscular bodies. Males are usually smaller than females, with a normal length range of two to six feet. Their bodies are shaped like cylinders and have glossy, smooth scales covering them.

- Conduct and Attitude

For reptile aficionados of all skill levels, king snakes are a popular pet despite their intimidating image as predators since they are typically gentle and easy to handle. King snakes can grow to be rather docile and even love being handled by their owners if they are given the right care and socialization from an early age.

King snakes are solitary animals that prefer to travel and hunt by themselves in their natural habitat. They are not territorial, though, and unless they are vying for food or mates, they will usually stay out of each other's way. King snakes can make protective movements such coiling, vibrating, or hissing when they feel threatened. They usually don't act aggressively, though, and will back down if given the chance.

- Enclosure and Housing Requirements

For the health and welfare of your king snake, you must provide the ideal habitat. It's critical to create an enclosure that as nearly resembles the snake's native habitat as possible. This entails offering enough of room for movement, a range of hiding places, and the right amount of humidity and temperature.

A king snake should be kept in a glass terrarium or plastic container with a tight-fitting lid. With ample space to spread out and explore, the enclosure should be big enough to fit the snake's size and activity level. The bottom of the box can be lined with a substrate, such as cypress mulch or aspen shavings, to give the snake a cozy place to hide and burrow.

The enclosure should have a range of enrichment materials, such as branches, rocks, and artificial plants, in addition to substrate. These will give the snake chances to climb, investigate, and hide, preventing boredom and encouraging natural behaviors.

- Temperature and Illumination Needs

For your king snake to be healthy and happy, its habitat needs to be kept at the right temperature and with the right amount of lighting. Since king snakes are ectothermic, their body temperature is controlled by outside heat sources. Because of this, it's critical to create a temperature gradient inside the enclosure so the snake can migrate between warmer and colder spots as needed.

To provide a comfortable basking area at one end of the enclosure, a heat source such as an under-tank heating pad or ceramic heat emitter should be included. The ideal range for daytime temperatures in this region is 85–90°F (29–32°C), with nighttime lows of 75–80°F (24–27°C).

King snakes need access to UVB lighting in addition to a heat source in order to properly metabolize calcium and vitamin D3, two critical nutrients for healthy bones. One end of the enclosure should have a fluorescent UVB bulb covering it so that it receives 10 to 12 hours of light per day.

- Nutrition and Feeding

A balanced diet is crucial for the health and wellbeing of your king snake. In the wild, small animals, snakes, and other reptiles are the main prey for king snakes. They can be fed rodents of the proper size, such as mice, rats, and chicks, while they are in captivity.

Adult king snakes can be fed every 7–10 days, however juveniles should only be fed every 5–7 days. To avoid choking or regurgitation, it is crucial to provide prey items that are approximately the same size as the thickest portion of the snake's body. It is best to use frozen-thawed prey items because they reduce the chance of giving your snake parasites or diseases.

- Managing and Introducing

The secret to building a solid relationship with your king snake is proper handling and socializing. It's crucial to handle your snake with composure and assurance, giving it time to get used to your presence. As you remove the snake from the enclosure, begin by softly caressing its body and work your way up to bearing its entire weight.

Handling your king snake right after eating or during shedding can stress it out and make it more likely for it to act defensively or aggressively. Rather, refrain from handling the snake until it has had a chance to finish shedding or digest its meal.

- Well-being and Health

It's critical to keep an eye on your king snake's health in order to identify and address any possible problems early on. Clear eyes, smooth scales, and a robust appetite are indicators of a healthy snake. It is imperative that you seek the advice of a veterinarian who specializes in reptile care if you observe any changes in your snake's appearance or behavior, including lethargy, hunger loss, or breathing difficulties.

Infections with mites, parasites, and respiratory tract infections are common health problems in king snakes. It is crucial to seek treatment as soon as a disease manifests itself because they can frequently be successfully treated with quick veterinarian care.

- Reproduction and Breeding

Reptile enthusiasts with experience may find breeding king snakes to be a fulfilling endeavor. Make sure that both of your snakes are in good health and condition before attempting to breed them. Slowly acclimate the male and female snakes while keeping a close eye on their behavior for indications of hostility or readiness for marriage.

The female snake will go through a 60–70 day gestation period after the successful mating event before laying her eggs. A good nesting box packed with moist substrate is what a gravid female needs in order to lay her eggs. The female will wrap her eggs around them to keep them warm and safe until they hatch.

Chapter 1

Selecting the Ideal King Snake: Types and Features

There are several things to think about before choosing a king snake as a pet. A thorough examination of the snake's characteristics is necessary to identify the ideal king snake, including its size, color, temperament, and maintenance needs. In order to assist you in selecting your scaly friend, we will examine the various forms and traits of king snakes in this extensive guide.

Numerous King Snake Varieties

The genus Lampropeltis, which has various species and subspecies, is home to king snakes. The size, color, and preferred environment of each kind of king snake are distinctive features. Among the most well-liked kinds of king snakes are the following:

The California King Snake, or Lampropeltis californiae, is a native of the western United States and is regarded as the most recognizable and iconic species of kingsnake. Though there are also frequent varieties with red, yellow, or orange bands, it is most recognized for its striking black and white banding.

The highly adaptable eastern king snake, or Lampropeltis getula, is found throughout the eastern United States and can be found in a range of environments, including grasslands, marshes, and woodlands. Though there are variants with complete black or white coloring as well, it is usually black with bands of yellow or white.

As its name implies, the Mexican Black KingSnake (Lampropeltis nigrita) is distinguished by its full black coloring. This species, which is native to Mexico, is highly valued by collectors because of its elegant look and gentle nature.

The desert king snake, or Lampropeltis splendida, is a species of snake that lives in dry desert regions and is found in northern Mexico and the southwestern United States. Its darker brown or black bands, which are usually tan or beige, let it blend in perfectly with its natural surroundings.

The Florida King Snake (Lampropeltis floridana) is a state-endemic snake that is distinguished by its vivid stripes of red, yellow, and black. Because of its similar coloring, people frequently confuse it for the deadly coral snake, even though it is not harmful to people.

These are just a handful of the several species of king snakes that aficionados of reptiles can choose from.

Before choosing a variety, it's crucial to conduct extensive research because each has distinct qualities and maintenance needs of its own.

Features of the King Snake

Apart from their diverse hues and designs, king snakes have several other attributes that set them apart from other types of reptiles. The following are a few of the most prominent traits of king snakes:

Constricting Prey: King snakes are strong constrictors, just as the other members of the Lampropeltis genus. They ensnare their prey by encircling it with their bodies and applying pressure until it suffocates. They can feed on a broad range of creatures, such as lizards, snakes, and small mammals, thanks to this adaptability.

Calm Temperament: King snakes are popular pets for reptile enthusiasts of all skill levels since they are often calm and easy to manage. King snakes can grow to be rather docile and even love being handled by their owners if they are given the right care and socialization from an early age.

Mimicry: Some king snake kinds, like the Florida and California king snakes, mimic venomous species to ward off predators. This behavior is known as Batesian mimicry. Their colors and banding patterns, which closely mimic those of venomous coral snakes, are especially indicative of this.

Hibernation: King snakes hibernate for a portion of the winter in cooler climates. They look for safe havens, such as underground tunnels, where they can hibernate until springtime temperatures rise. Their metabolic rate drastically decreases during hibernation, enabling them to store energy until food becomes more plentiful.

Longevity: King snakes in captivity can survive for up to fifteen years or longer if given the right care. Making sure your snake has a long and healthy life requires frequent veterinary care, a balanced diet, and a proper habitat.

Selecting the Ideal King Snake for Your Needs

It's critical to take into account your lifestyle, tastes, and expertise level before choosing a king snake as a pet. While some king snake varieties might be easier for novices to handle, others might need more sophisticated care. In addition, considerations including hue, temperament, and size should be made.

A calm and manageable kind of king snake, such as the Mexican black king snake or the California king snake, can be the ideal option for you if you're new to raising reptiles. These species are perfect for first-time keepers because they are often more tolerant of small care and handling errors.

However, if you're an experienced reptile lover seeking a challenge, you might be more interested in the more specialist king snake kinds, like the Florida king snake or the desert king snake. Even though some animals may require more specialized care and have peculiar temperaments, people who are prepared to put in the extra work can reap great rewards from owning one of these species.

Whichever variety you decide on, it's crucial to do extensive study and speak with knowledgeable keepers or breeders before committing. You can make sure you get the ideal scaly family member by taking the time to educate yourself about the various types and traits of king snakes.

Chapter 2

Setting up the Perfect Environment and Housing

Creating a proper home for your king snake is essential to its general health, happiness, and well-being. In addition to offering your snake a cozy place to live, a well-designed enclosure will help to mimic its natural habitat, enabling it to display its typical habits and flourish in captivity. We will go over all the necessary components of setting up the perfect environment for your king snake in this extensive guide, including enclosure selection, substrate selection, controlling humidity and temperature, décor options, and more.

Choice of Enclosure

The first step in setting up your king snake for optimal habitat is to choose a suitable enclosure. When selecting an enclosure, there are a few things to take into account: size, material, and design. Here are a few typical choices to think about:

Glass Terrariums: Because they are transparent and require little upkeep, glass terrariums are a common option for housing king snakes. They give you an unobstructed view of the snake and its surroundings, making it simple to keep an eye on its activity and well-being. Glass terrariums can accommodate snakes of various sizes because they are available in an assortment of shapes and sizes.

Plastic Enclosures: Another great choice for sheltering king snakes are plastic enclosures, like those composed of HDPE or PVC. They are perfect for both novice and seasoned keepers because they are lightweight, strong, and simple to clean. Additionally, plastic cages are good at retaining humidity and heat, giving your snake a consistent habitat.

Wooden Vivariums: When building realistic environments for king snakes, wooden vivariums are a common option. They can be adorned with a range of décor elements, such as branches, rocks, and plants, and offer exceptional insulation. Additionally, wooden vivariums hold heat and humidity effectively, which makes them appropriate for species that need greater humidity levels.

No matter what kind of enclosure you select, you must make sure that it has enough ventilation to keep moisture and stagnant air from building up. Screen tops and mesh panels are frequently utilized to provide ventilation without sacrificing security.

Size of Enclosure

It's crucial to pick an enclosure for your king snake that gives it plenty of room to walk around and explore.

Generally speaking, you should have an enclosure that is at least twice as long as the snake's body and between one-third and half as wide.

A four-foot-long king snake, for instance, would need an enclosure that measures at least eight feet in length, two feet in width, and one to two feet in height. Giving the snake lots of space to spread out and explore will reduce stress and encourage natural activities.

Selection of Substance

Selecting the appropriate substrate for your king snake's habitat is crucial to preserving good health and encouraging organic habits. There are various substrate choices available, and each has benefits and drawbacks of its own:

Aspen Shavings: Because of their low cost, high absorbency, and simplicity of maintenance, aspen shavings are a well-liked substrate option for king snakes. They assist in controlling the humidity levels in the enclosure and give the snake a cozy place to hide and burrow.

Cypress Mulch: For king snakes, cypress mulch is an additional great substrate choice. Because it is natural, absorbs moisture, and suppresses odors, it is perfect for high-humidity settings. In addition to having a realistic appearance, cypress mulch is spot-cleanable.

Reptile Carpet: Made of sturdy synthetic materials like nylon or polyester, reptile carpet is a synthetic substrate. If swallowed by the snake, it aids in preventing impaction and is reusable and easily cleaned. Reptile carpet, however, could not offer as realistic of a setting as alternative substrate choices.

Paper Towels/Newspaper: For young or unwell snakes, paper towels or newspaper are good temporary substrate solutions. They are perfect in circumstances when hygiene is of the utmost importance because they are affordable, disposable, and simple to replace.

Control of Temperature and Humidity

For your king snake to remain healthy and happy, its habitat needs to be kept at the right temperature and humidity conditions. Since king snakes are ectothermic, their body temperature is controlled by outside heat sources. It is crucial to create a thermal gradient in the enclosure so the snake can migrate between places that are warmer and cooler as needed.

One end of the enclosure should have a basking area that is 85–90°F (29–32°C), while the other end should be somewhat colder, at 75–80°F (24–27°C). The snake will be able to efficiently thermoregulate and satisfy its metabolic requirements thanks to this gradient.

King snakes need heat, but they also need the right amount of humidity to keep their skin and respiratory system in good condition. For the most part, king snake species should have between 50 and 60 percent humidity; however, some species may need more humidity, particularly during shedding.

Provide a shallow water dish for the snake to soak in and spritz the enclosure with water as needed to control the humidity levels. Increasing the amount of humidity in the enclosure and giving your snake a more realistic habitat can also be achieved by adding real or artificial plants.

Accents and Enhancements

In addition to being aesthetically beautiful, adding naturalistic décor and enrichment items to your king snake's habitat will encourage natural behaviors and lessen stress. Consider the following décor options:

Branches and Rocks: Including branches and rocks in the cage gives your snake extra places to climb and explore, as well as a more lively atmosphere. Select robust branches and rocks that are firmly fixed to avoid them from moving or toppling over.

Hides & Shelters: To give your king snake a feeling of security and seclusion, hides and shelters are crucial. To let the snake select its favorite hiding place, place several hides throughout the cage, one on the warm end and one on the chilly end.

Live or Artificial Plants: Adding live or artificial plants to the enclosure raises the humidity levels and gives your snake more stimulation in addition to making it look better. Make sure the plants you choose are safe for reptiles and are firmly anchored to keep them from being uprooted.

Environmental Enrichment: To keep your king snake psychologically engaged and active, you must provide it with environmental enrichment. Give your snake the chance to hunt, explore, and exhibit natural behaviors. Some examples of these activities include hiding food in various parts of the cage or offering real prey items.

You may build a fascinating and engaging habitat for your king snake that encourages natural behaviors and improves the snake's overall quality of life by adding these components to it.

Upkeep and Sanitization

Maintaining a clean and hygienic enclosure for your king snake requires routine upkeep and cleaning. The following advice will help you keep your snake's habitat tidy and sanitary:

Spot-clean the enclosure every day to get rid of filthy substrate, shed skin, and excrement. To stop the growth of bacteria and odor, replace any dirty substrate with new material as needed.

At least once a month, give the enclosure a thorough cleaning. This involves taking out all of the décor and substrate, cleaning the space with a cleaner appropriate for reptiles, and adding new substrate and décor.

Water dishes, hides, and other décor objects should be cleaned and disinfected on a regular basis to stop the growth of mold and bacteria. After giving them a good rinse in hot water, let them air dry fully before putting them back in the enclosure.

To make sure that the temperature and humidity levels stay within the proper range for the species of your snake, frequently check them and make any necessary adjustments.

You can make sure that your king snake's habitat stays tidy, sanitary, and supportive of its general health and well-being by paying attention to these maintenance guidelines.

Significant thought must go into selecting the right enclosure, substrate, controlling humidity and temperature, choosing décor, and maintaining and cleaning procedures in order to create the perfect home for your king snake. You may support the physical and emotional well-being of the snake and guarantee that it has a happy and meaningful existence in captivity by creating a cozy and stimulating environment that closely resembles its native home. You and your king snake can enjoy your newly created home for many years to come with the right maintenance and attention to detail.

Chapter 3

Feeding Your King Snake: Nutritional and Dietary Advice

Your king snake's health and wellbeing depend on proper feeding. King snakes are opportunistic predators that have adapted to eat a variety of animals, mostly other reptiles like lizards, snakes, and occasionally small mammals. To make sure your snake gets the nutrition it needs to flourish in captivity, it's critical to mimic this natural diet as nearly as possible. We will examine the food needs of king snakes in this extensive guide, including topics such as feeding frequency, prey selection, prey size, and supplements.

Frequency of Feeding

The amount, size, and activity level of your king snake determine how often you should feed it. Because they are still growing quickly, juvenile king snakes usually need to be fed more frequently than adult snakes. Juvenile king snakes should generally be fed every 5-7 days, however adult king snakes can be fed every 7–10 days.

It's critical to keep an eye on your snake's physical condition and modify the frequency of feedings as necessary. Your snake may require more frequent feedings to support its growth if it is expanding quickly while keeping a healthy body weight. On the other hand, you might need to lessen the frequency of feedings or change the size of the prey items if your snake starts to gain weight or become obese.

Selection of Prey

King snakes hunt mostly other reptiles in the wild, such as lizards and snakes, and sometimes small mammals like rats. To make sure your king snake gets the nutrition it needs to flourish in captivity, it's critical to provide it a meal that closely resembles its native diet.

Mice and rats are among the easiest and most accessible prey items for king snakes kept in captivity. These rodents are widely accessible from breeders and pet retailers, and they are available in a range of sizes to accommodate snakes of all shapes and sizes. To avoid choking or regurgitation, it's crucial to make sure the prey items you provide your snake are the right size.

To provide your snake diversity and enrichment, you can offer several kinds of prey in addition to mice and rats. The following are a few things that king snakes can feed on:

Chicks: King snakes may easily find and consume day-old chicks as a wholesome source of prey. They are a great option for snakes that need extra calories to support development or reproduction because they are high in fat and protein.

Quail: King snakes can also effectively feed on quail. Even though they are smaller than chicks, they are still a fantastic source of fat and protein for your snake.

Other Reptiles: You can provide your king snake with prey other than rodents and birds, such as different kinds of reptiles. You can occasionally provide your snake small lizards for diversity and enrichment, like geckos or anoles.

To avoid introducing parasites or disease, be sure that any prey items you feed are the right size for your snake and come from a reliable source.

Size of Prey

It is crucial that you choose prey items for your king snake that are suitable in size for both its age and size. Prey items that are too big could choke your snake or make it regurgitate its food; on the other hand, prey items that are too little might not give your snake enough nutrients to meet its demands.

Prey objects should, in general, be about the same size as the thickest portion of your snake's body. This may entail beginning with smaller prey items, like fuzzy or pinky mice, for juvenile king snakes and progressively increasing the size as the snake matures. To suit their dietary needs, adult king snakes can be given larger prey items, including adult mice or small rats.

It's critical to keep an eye on your snake's physical health and to modify the size of its prey items appropriately. To assist your snake maintain a healthy body weight, you may need to decrease the size of the prey items or lengthen the intervals between feedings if it is getting overweight or obese.

Addendum

To make sure your king snake gets all the vital vitamins and minerals it needs to grow, supplementation could be required in addition to feeding it a varied and balanced diet. For the health of their bones and metabolism, reptiles require a lot of calcium and vitamin D3, which may need to be supplied if their diet and exposure to UVB rays are insufficient.

Before giving prey items to your snake, one of the most popular ways to supplement king snakes is to dust them with a calcium powder that contains vitamin D3. This makes it more likely that your snake will get the nutrients it needs to finish its meal. A UVB light source placed within the enclosure can also aid in your snake's better metabolism of calcium and vitamin D3.

It's critical to seek advice from a qualified veterinarian or knowledgeable reptile keeper to ascertain your king snake's proper supplementing schedule depending on factors including size, age, and nutritional requirements. It's crucial to pay great attention to your snake's health and strictly adhere to dose recommendations as over-supplementation may be detrimental to it.

Feeding Techniques

There are various ways to feed your king snake, and each has benefits and drawbacks of its own. The size, temperament, and preferred feeding schedule of your snake will determine which method you use. King snakes frequently use the following as food sources:

Using feeding tongs or forceps, hand feed your snake by holding objects that would normally be considered prey. You may watch your snake's feeding habits and engage in intimate interaction with it using this method. For shy or reluctant to eat snakes, hand feeding can be quite helpful.

Bowl feeding entails putting prey items within the snake's enclosure in a shallow dish or bowl. Using this technique, your snake can hunt and eat at its own speed, just like it would in the wild. For snakes that become agitated or stressed out easily, bowl feeding may be beneficial.

Allowing your snake to hunt and eat objects that are placed just within the enclosure is known as "free feeding." Your snake will benefit from this method's enrichment and cerebral stimulation, but if it doesn't finish all of the prey, there may be an excess of food waste.

Whichever feeding technique you decide on, it's critical to keep a close eye on your snake's feeding habits and make any adjustments to guarantee it gets the right amount of food and nourishment.

Hints and Advice on Feeding

There are a few things to remember and guidelines to follow when feeding your king snake in order to have a good and safe feeding experience:

Feed in a Separate Enclosure: It is advisable to feed your snake in a separate enclosure that has been set aside just for that purpose in order to lower the possibility of unintentional ingestion of substrate or other foreign materials. Additionally, by doing this, you can lessen the chance of an unintentional bite during handling by keeping your snake from confusing your hand with food.

After feeding, keep a close eye on your snake to make sure it is appropriately breaking down its food. After feeding, a small bulge in the snake's abdomen is typical; however, you should contact a veterinarian right once if you observe any symptoms of regurgitation, bloating, or discomfort.

Maintain Hygiene: To stop the growth of mold and germs, keep water and feeding dishes clean and sanitized. To avoid spoiling and stench, remove any uneaten prey items from the enclosure as soon as possible.

Avoid Handling Your Snake just After Feeding: Hold off on handling your snake just after feeding to minimize stress and the chance of regurgitation. Before you handle or contact your snake, give it some time to finish its meal without being disturbed.

You can make sure that your king snake has the proper diet and care to flourish in captivity by following these rules and considerations. Your snake will be a valued member of your family and live a happy and healthy life if given the right care and diet.

Chapter 4

Taking Care When Handling KingSnakes: Safe Handling Practices

Getting to know your king snake and witnessing its amazing actions up close can be both rewarding and a bonding experience. To protect your health and the health of your pet, you must handle your snake sensibly and safely. We will go over safe handling guidelines, how to handle and pick up your king snake, how to recognize symptoms of stress or pain, and how to develop a relationship of trust and confidence with your snake in this extensive book.

Guidelines for Safe Handling

To avoid hurting yourself or your pet, it's important to know the fundamentals of safe handling before attempting to handle your king snake. The following general rules should be remembered:

Handle Your Snake Care: Always wash your hands well with soap and water before and after touching your snake. By doing this, you and your snake can avoid transferring bacteria and other infections to one another.

Handle Gently: Take care when handling your snake to prevent hurting or upsetting it. Squeezing or pinching the snake should be avoided as this may irritate it or make it defensive.

Support the Body: To minimize harm and increase comfort, always provide your snake with the right support when you pick it up and hold it. Your snake's spine may get harmed if you take it up by its tail.

Avoid Quick Movements: When handling your snake, try not to move quickly or jerkily as this may frighten or agitate the creature. To make your snake feel comfortable and protected, move gently and deliberately.

Children Under Close Supervision: To avoid mishaps or injuries, ensure that children handling snakes are properly taught safe handling practices and are constantly under close supervision.

Know Your Snake: Before attempting to handle your snake, familiarize yourself with its habits and disposition. It's crucial to determine your snake's comfort level and make adjustments based on it, as certain snakes may be more amenable to handling than others.

Grasping and Maintaining Your Snake

To help your king snake feel safe and secure, approach gently and confidently when taking it up and holding it. The following are detailed guidelines for managing your snake safely:

Approach Calmly: To avoid frightening or upsetting your snake, approach it with confidence and without making any abrupt movements or loud noises.

Support the Body: Carefully lift your snake from underneath with both hands, making sure to support its entire body—from head to tail. It is best not to grab or squeeze the snake since this may make it uncomfortable or make it act defensively.

Lift Gently: Raise your snake gently and gradually so that it can become used to being handled. To avoid frightening or upsetting the snake, do not lift it abruptly or violently.

Hold Firmly: After you have raised your snake, use both hands to support its body as you hold it firmly but softly. Refrain from holding on to the snake too tightly as this may result in pain or damage.

Watch for Signs of Discomfiture: Keep a close eye out for any indications that your snake is stressed or uncomfortable, such as hissing, striking, or making an attempt to flee. Try again later if your snake exhibits any indications of distress, and carefully lower it back into its container.

Keep Handling Sessions Brief: To give your snake time to get used to being handled, keep handling sessions brief, especially in the beginning. As your snake gets more at ease, gradually extend the length and frequency of handling sessions.

Indices of Unease or Stress

To avoid harm or suffering, it's critical to keep a close eye out for any indications of tension or pain when handling your snake. Among the typical indicators of stress in king snakes are:

Hissing is a defensive activity that snakes frequently display when they are anxious or feel threatened. Your snake may hiss while it is being handled because it may be feeling uneasy or insecure.

Another defensive activity that snakes may do when they feel agitated or threatened is striking. It is imperative that you lower your snake back into its enclosure and give it some time to calm down if it attacks you while being handled.

Tail Vibrations: When under stress or feeling threatened, certain snakes may shake their tails quickly. This action indicates that the snake is defensive and is frequently followed by hissing or striking.

Trying to Get Away: If your snake tries to get away from you while you are handling it, it can be indicating that it is anxious or uneasy. To avoid harm or escape, you must treat your snake with care and security.

Defensive Posture: In snakes, defensive postures such as coiling or flattening the body are frequently indicative of stress or discomfort. Lower your snake back into its enclosure and allow it some time to unwind if it displays defensive postures during handling.

It is imperative that you quickly drop your snake back into its enclosure and give it some time to settle down if you observe any of these symptoms of stress or pain in it during handling. Short, gentle handling sessions are best to avoid overwhelming or overstressing the animal.

Developing Self-Belief and Trust

It takes time and care to establish trust and confidence with your king snake, but it's necessary to forge a solid link and foster a happy relationship. Here are some pointers for fostering confidence and trust in your snake:

Handle Often: Handle your snake frequently to help it get used to being handled and gradually develop confidence and trust. As your snake gets more comfortable, start with brief handling sessions and progressively extend them in length and frequency.

Be Patient: Especially if your snake is timid or anxious, show it tolerance and patience. The handling procedure should not be rushed or forced since this can lead to tension and impede progress.

Use Positive Reinforcement: To help your snake trust and engage with you, use positive reinforcement techniques like giving it food rewards or giving it soft praise. Associate handling with pleasant memories to increase your snake's sense of security and comfort.

Respect bounds: Don't push your snake past its comfort zone; instead, pay attention to its cues and bounds. Give your snake some room and try again later if it exhibits indications of stress or pain.

Be constant: To make your snake feel safe and enable you to anticipate its interactions with it, be constant in your handling style and regimen. Over time, establishing a consistent handling plan might aid in the development of confidence and trust.

You can help your king snake feel comfortable and secure when being handled and foster a strong bond and positive relationship between it and you as its owner by using these helpful hints and tactics.

Getting to know your king snake and witnessing its amazing actions up close may be both rewarding and pleasurable experiences. You can make sure that your interactions with your snake are beneficial and enriching for both of you by adhering to the principles of safe handling, using the right techniques for picking up and holding your snake, keeping an eye out for any signs of stress or discomfort, and gradually developing trust and confidence. You can create a strong link and a long-lasting relationship with your king snake that will enrich and fulfill both of your lives if you are patient, consistent, and take care of it.

Chapter 5

Health and Well-Being: Typical Problems and Appropriate Treatment for King Snakes

For your king snake to have a long and healthy life, you must take care of its health and wellbeing. King snakes are prone to a wide range of health problems, from simple illnesses to more serious disorders, much like all other reptiles. We will go over frequent health problems affecting king snakes, warning signs and symptoms, preventive care options, and appropriate veterinary care procedures in this extensive guide.

Typical Health Concerns

Even though they are tough and durable in general, king snakes can nevertheless develop a number of health problems, some of which are more prevalent than others. Early detection and treatment of many frequent health disorders depend on an understanding of the signs and symptoms that correspond with them. The following are a few of the most typical health problems king snakes face:

One of the most prevalent medical conditions affecting captive king snakes is respiratory illness. These illnesses are frequently brought on by poor husbandry techniques, such as low humidity or temperature, and can be brought on by bacteria, viruses, or parasites. King snake respiratory infections can cause wheezing, difficult breathing, nasal discharge, and lethargic behavior.

Outside Parasites: If left untreated, external parasites like mites and ticks can infest king snakes, causing irritation, discomfort, and even secondary diseases. The substrate, décor, or prey items that are contaminated are common ways for these parasites to enter the cage.

King snakes may exhibit visible parasites on their skin, frequent rubbing or scratching of surfaces, and skin blisters or irritation as indicators of external parasites.

Internal Parasites: King snakes are susceptible to internal parasite infections in their gastrointestinal tract, which can cause symptoms including diarrhea, weight loss, lethargy, and poor appetite. These parasites include nematodes, coccidia, and protozoa. Usually, consuming contaminated food or water supplies is how these parasites are acquired.

Skin Infections: Trauma, damage, and bacterial or fungal infections can all lead to skin infections in king snakes, including abscesses and ulcerations. Veterinarian intervention may be necessary to treat these infections, which might manifest as localized swelling, redness, or discharge on the skin.

Metabolic Bone Disease (MBD): Inadequate consumption of calcium and vitamin D3 is the primary cause of this nutritional condition, which is frequently observed in captive reptiles, particularly king snakes. Weakened bones, abnormalities, and neurological symptoms like seizures or tremors can all result from multiple sclerosis. In order to manage MBD in king snakes, prevention and early intervention are crucial.

Preventive Healthcare Practices

King snake health and wellbeing are greatly dependent on preventive care, which also lowers the likelihood of frequent health problems. You should take the following precautions for your king snake:

Provide Appropriate Husbandry: To provide a cozy and stress-free atmosphere, keep your king snake's habitat at the right temperature, humidity, and lighting levels.

To make sure these parameters stay within the proper range for the species of your snake, keep a close eye on them and adjust as necessary.

Provide a Variety and Balanced Diet: To guarantee that your king snake gets all the vital nutrients it needs to flourish, provide a varied and balanced diet made up of suitably sized prey items. When necessary, add calcium and vitamin D3 supplements to prey foods to help prevent nutritional deficits.

Before adding any new reptiles or items of prey to your king snake's habitat, quarantine them to stop the spread of parasites or contagious diseases. Before permitting anyone in quarantine to engage with your snake, keep a tight eye out for any indications of disease or infestation.

Maintain Good Hygiene: To avoid the growth of germs, parasites, and other pathogens, keep the enclosure housing your king snake clean and sanitized. Feces, shed skin, and uneaten prey items should be removed right away. Surfaces should also be routinely cleaned using a cleaner safe for reptiles.

Handle with Care: To prevent stress or harm, handle your king snake sensibly and gently. Before and after touching your snake, wash your hands to stop the spread of germs and other illnesses.

Plan Frequent Veterinary Check-ups: To keep an eye on your king snake's health and identify any possible problems early, schedule routine veterinary examinations. To make sure your snake stays in good condition, a knowledgeable veterinarian with experience with reptiles can do diagnostic testing, a complete physical examination, and a fecal study.

Appropriate Veterinary Treatment

For any health problems that may develop in your king snake, appropriate veterinary care is crucial in addition to preventive care measures. It's critical to get veterinarian care as soon as you see any indications of disease or damage in your snake in order to stop the situation from getting worse. The following recommendations are for giving king snakes the right veterinary care:

Locate a Veterinarian with Reptile-Savvy Experience: To take care of your king snake, seek out a veterinarian with experience and knowledge in reptile medicine. Because of their distinct anatomical and physiological traits, reptiles need specific knowledge and training in order to diagnose and treat them.

Plan Regular Check-ups: To keep an eye on your king snake's general health and identify any possible problems early, schedule routine veterinary examinations. Your veterinarian can evaluate the health of your snake at these check-ups by doing a complete physical examination, analyzing its feces, and ordering any necessary diagnostic tests.

Seek Quick Treatment: It's critical to get veterinarian help as soon as you observe any indications of disease or injury in your king snake, such as behavioral, dietary, or physical changes. Prompt action will help keep your snake's condition from getting worse and improve its outlook.

Treatment instructions: To guarantee the best possible outcome for your king snake, closely adhere to your veterinarian's treatment instructions. This could entail giving your snake supportive care, giving medication, and modifying its diet or husbandry as necessary.

Watch Recovery: Throughout the healing phase, keep a close eye on your king snake and follow up with your veterinarian as instructed. Any changes in your snake's health, habits, or hunger should be noted and immediately reported to your veterinarian.

You can make sure your king snake gets the right care and support to get over any health problems it may have by adhering to these instructions for adequate veterinary care.

To guarantee the health and welfare of your king snake, you must be diligent, watchful, and provide the right care. You may contribute to your snake's happy and healthy life in captivity by being aware of frequent health problems that affect king snakes, putting preventive care measures in place, and getting quick veterinary intervention when necessary. Your king snake can provide you and your family with many years of companionship and delight if you provide it with the right diet, care, and veterinarian attention.

Chapter 5

Sophisticated Maintenance Advice for Skilled King Snake Keepers

Learning more about advanced care procedures will help keepers of king snakes who are already experienced improve the health, overall quality of life, and overall well-being of these interesting reptiles. Comprehending the nuances of king snake husbandry, behavior, enrichment, breeding, and medical treatment is necessary for advanced care. We will go over advanced care methods and strategies in this extensive book for knowledgeable keepers to improve the upkeep and enjoyment of their king snakes.

Advanced Methods of Husbandry

Beyond simply giving king snakes food, water, and shelter, advanced husbandry techniques concentrate on establishing ideal living conditions that support their natural activities, health, and well-being. Here are some sophisticated husbandry methods for knowledgeable custodians:

Environmental Enrichment: Add climbing branches, hiding places, and sensory stimuli to your king snake's habitat to make it more appealing to it. Give your snake the chance to hunt, explore, and participate in its natural activities to keep it mentally and physically stimulated.

Bioactive Substrate: To give your king snake a more realistic and self-sustaining habitat, think about utilizing a bioactive substrate, like a mixture of organic soil, sphagnum moss, and leaf litter. Beneficial microorganisms are fostered and a dynamic ecosystem is created by bioactive substrates, which contribute to the cleanliness and health of the soil.

Naturalistic décor: To create a visually pleasing and enriching habitat, add naturalistic décor pieces to your king snake's enclosure, such as driftwood, rocks, and real or artificial plants. Select décor pieces that offer chances for exploration and shelter while paying homage to the natural habitat of the snake.

Personalized Lighting: Tailor the lighting arrangement for your king snake to resemble the photoperiod and intensity of its natural habitat. To enhance the production of vitamin D and the metabolism of calcium, use full-spectrum UVB lighting. To replicate natural lighting, create a day-night cycle with gentle transitions.

Seasonal Variation: Throughout the year, change the temperature, humidity, and photoperiod to provide your king snake with a more varied habitat. Encourage natural breeding practices and reproductive cycles by modeling seasonal variations in temperature and sunshine hours.

Behavioral Monitoring and Interpretation

Expert king snake keepers recognize how critical it is to watch and analyze their snake's behavior in order to precisely gauge its general health, emotional state, and general well-being. Gaining a better comprehension of your snake's behavior can help you better predict its requirements, inclinations, and reactions. The following advice can help you observe and comprehend the behavior of king snakes:

Body Language: To effectively read your king snake's attitude and intents, pay attention to its posture, movements, and body language. Learn to identify subtle clues, like body language, tongue flicking, and vocalizations, that indicate stress, hostility, curiosity, or contentment.

Feeding Behavior: To gauge your king snake's appetite, feeding reaction, and preferred prey, watch how it feeds. Take note of any changes in feeding patterns, such as excessive eating, regurgitation, or reluctance to eat, as these could be signs of underlying medical conditions or stressors in the environment.

Social Interactions: If you have more than one king snake or other reptile, keep an eye on how they interact with each other to keep an eye on dominance, hierarchy, and territorial behavior. Observe any indications of hostility, rivalry, or tension among the group and modify the enclosure configuration or group configuration as necessary.

Reproductive activity: Keep an eye out for indicators of reproductive activity in your king snakes throughout mating season, such as courtship displays, mating rituals, and female ovulation. To promote successful mating and egg laying, provide suitable nesting places, temperature gradients, and photoperiod cues.

Environmental Responses: Pay attention to how your king snake reacts to variations in its surroundings, including temperature swings, humidity levels, and illumination. Take note of any alterations in behavior or inclinations towards particular environmental conditions and modify the enclosure parameters correspondingly.

Improvement and Mental Excitation

In order to keep king snakes in captivity psychologically stimulated, engaged, and fulfilled, enrichment activities are essential. Expert king snake keepers use a range of enrichment methods to encourage natural activities, avoid boredom, and improve general health. For seasoned keepers, consider these enrichment suggestions:

Foraging Opportunities: By concealing prey items or scent trails within the cage, you can provide your king snake the chance to engage in natural foraging activities. To promote hunting and exploration, use interactive feeders, smell enrichment, and feeding puzzles.

Environmental Manipulation: To give your king snake lively and captivating surroundings, play about with the substrate, décor, and temperature gradients. To arouse the senses and encourage movement, present novel textures, smells, and challenges.

Behavioral Training: Use positive reinforcement methods to teach your king snake basic behaviors including recall, stationing, and target training. Establish trust and communication with your snake by rewarding desired behaviors with food, using clicker training, or using tactile cues.

Environmental Modifications: To avoid habituation and promote curiosity and adaptability, periodically introduce modifications to the enclosure's design, furnishings, or enrichment materials. To encourage curiosity and mental engagement, move hiding places, adjust décor, or add interesting objects.

Sensory Enrichment: Engage your king snake in tactile, visual, aural, and olfactory enrichment activities to stimulate its senses. Provide visual stimuli like mirrored surfaces or live prey, offer scented objects or pheromone cues, play relaxing nature sounds, and change the textures of the substrate to encourage tactile exploration.

Breeding and Taking Care of Reproduction

Understanding the nuances of reproductive biology, breeding behavior, and neonatal care is crucial for experienced keepers who wish to successfully produce king snakes. For seasoned keepers, consider the following advanced breeding and reproductive care advice:

Seasonal Cycling: To encourage king snake reproduction and breeding activity, mimic seasonal variations in temperature, humidity, and photoperiod. Reduce wintertime temperatures and sunshine hours gradually to encourage brumation and get ready for breeding season.

Breeding strategies involve pairing up compatible male and female king snakes based on traits like size, temperament, age, and genetic compatibility. To evaluate the timing and success of breeding, keep an eye on mating activity, wooing displays, and copulation.

Nesting and Egg Laying: Provide suitable locations for gravid female king snakes to lay their eggs, such as wet substrate-filled humid hide boxes. Prior to the release of eggs, nesting places should be prepared, and females should be closely observed for indications of pre-ovulatory swelling, follicular expansion, and ovulation.

The best way to ensure that king snake eggs hatch is to incubate them in a controlled environment with consistent humidity and temperature. Throughout the incubation period, keep the conditions constant and use a suitable incubation medium, such as vermiculite or perlite.

Neonatal Care: Give newborn king snakes the attention and care they need, which includes keeping the right temperature and humidity conditions, providing suitable prey, and keeping an eye on their development. Reduce stress and handle newborns gently to encourage healthy development and acclimatization.

Medical Attention and Veterinary Assistance

Expert king snake keepers recognize the value of preventative medicine, regular health checks, and fast veterinary assistance when necessary. For seasoned keepers, consider these advanced medical care recommendations:

Regular Health Checks: Keep an eye on your king snakes' weight, skin tone, general health, and body condition by doing routine health checks. Maintain thorough records of all observations, feeding patterns, shedding schedules, and any modifications to your health.

Diagnostic Testing: To evaluate your health, screen for parasites or infectious diseases, and find underlying health issues early on, use diagnostic tests like fecal analysis, blood work, radiography, and ultrasound.

Preventive Medicine: To reduce the danger of infectious diseases and maintain your king snakes' best health, use preventive medicine techniques including vaccination, parasite management, and environmental hygiene.

First Aid and Emergency Care: In the event of an accident, disease, or medical emergency, be ready to provide your king snakes with first aid and emergency care. A well-stocked first aid kit should always be kept on hand, and you should become knowledgeable about basic first aid for reptiles, including supportive treatment, fluid therapy, and wound care.

Consultation with Reptile Veterinarian: Get in touch with a vet who understands reptiles and has experience treating and caring for king snakes. See your veterinarian on a regular basis for guidance on preventive care, health examinations, and treatment suggestions according to the needs of your snake.

Advanced care for king snakes entails a thorough grasp of behavior, enrichment, breeding, husbandry, and medical treatment methods to ensure the best possible health and well-being for these amazing reptiles. Expert keepers may raise their king snakes' quality of life and increase their admiration for these amazing animals by employing cutting-edge care techniques. King snakes kept in captivity can receive the best care and stewardship from professional keepers who possess the necessary skills, knowledge, and passion to excellence.

Chapter 6

Reproduction and Breeding Considerations in the Breeding of King Snakes

For reptile lovers who are concerned in conservation, genetic variety, and the preservation of species, breeding king snakes can be a fulfilling and exciting undertaking. But careful planning, preparation, and knowledge of the reproductive biology and breeding habits of king snakes are necessary for effective reproduction. The reproductive anatomy of king snakes, breeding considerations, mating behavior, egg incubation, neonatal care, and ethical considerations for appropriate breeding techniques are all covered in this extensive book.

Anatomy of Reproduction

Monitoring reproductive health and ensuring successful breeding are dependent on an understanding of the architecture of the king snake's reproductive system. Like all other reptiles, king snakes have internal reproductive systems that are designed specifically for mating, fertilization, and laying eggs. King snakes' reproductive anatomy consists of the following:

Testes: The paired testes of male king snakes are situated inside the body cavity, close to the kidneys. During copulation, sperm cells, or spermatozoa, are transferred from the testes to the female reproductive canal.

Ovaries: The paired ovaries of female king snakes are found inside the body cavity, close to the kidneys. The process of embryonic development begins with the fertilization of eggs (ova) by sperm cells in the ovaries.

Oviducts: In female king snakes, the oviducts are long, coiled tubes that connect the ovaries to the cloaca. Within the oviducts, where they are nourished and enveloped in protective membranes, the eggs undergo fertilization in preparation for laying.

Hemipenes: Specialized copulatory organs found inside the cloaca, hemipenes are paired in male king snakes. To transfer sperm for fertilization, the hemipenes are everted, or turned inside out, and inserted into the female's cloaca during mating.

Breeding-Related Issues

To guarantee the health, welfare, and success of the breeding project, it is imperative to take into account a number of things prior to attempting to breed king snakes. King snake breeding considerations include:

Species Selection: Based on traits like age, size, temperament, and genetic variety, choose mateable male and female king snakes for breeding. Steer clear of breeding closely related individuals to reduce the likelihood of genetic abnormalities and to encourage general health and vitality in the progeny.

Before reproducing, make sure the male and female king snakes are in good health and condition. Assess reproductive fitness and find any underlying health issues that could impact breeding success by doing routine health checks, such as physical exams, fecal analyses, and parasite screening.

Environmental Preparation: Establish the right lighting, humidity, temperature, and enclosure layout to provide king snakes with a good breeding habitat. Simulate seasonal variations in temperature and photoperiod to promote spawning and reproductive activities.

Pre-Breeding Conditioning: To replicate winter conditions and cause brumation, gradually lower temperatures and daylight hours for breeding pairs of king snakes. To guarantee optimum health and reproductive preparedness, give plenty of food and water throughout the pre-breeding phase.

Breeding Compatibility: To ensure successful breeding, introduce a compatible pair of male and female king snakes. Carefully observe their interactions for indications of courtship, mating activity, and copulation. To stop hostility or tension and to prevent harm during mating efforts, separate incompatible pairs.

Mating Patterns

King snakes engage in intricate copulatory rituals, tactile interactions, and wooing displays as part of their mating activity. Hormonal fluctuations, reproductive preparedness, and environmental cues all affect king snake mating behavior. The following are some typical mating habits of king snakes:

Courtship Displays: To entice female partners and maintain dominance, male king snakes frequently put on lavish courtship displays. To indicate that an animal is ready to mate, courtship displays can include body undulations, chin rubbing, tail vibrations, and pheromone releases.

Tactile Interactions: To promote receptivity and copulation, male king snakes may engage in tactile interactions with their female partners during courtship, such as body stroking, chin resting, and head bobbing.

King snake copulation rituals usually include the male mounting the female from behind and lining up his cloaca with hers to allow for the transfer of sperm.

Before effective copulation takes place, the male may probe and stimulate the female's cloaca with his hemipenes.

Post-Copulatory activities: To strengthen pair bonding and increase the likelihood of successful reproduction, male and female king snakes may display post-copulatory activities such as mutual grooming, coiling together, or relaxing nearby.

Incubation of Eggs

Female king snakes go through gestation and egg development after a successful copulation, and then they lay a clutch of eggs. In order to ensure embryo viability and successful hatching, egg incubation is a crucial step in the reproductive process that requires precise temperature and humidity control. Some things to think about when incubating eggs in king snakes are as follows:

Choose a Nesting Site: Give gravid female king snakes an appropriate place to lay their eggs, like a humid hide box filled with wet substrate. To reduce disturbance and stress, select a nesting place in a peaceful, isolated part of the enclosure.

Egg Laying Behavior: Keep a watchful eye out for pre-ovulatory swelling, follicular expansion, and ovulation, which are indicators of upcoming egg laying in gravid female king snakes. Give the female enough space and nesting materials to lay her eggs peacefully and undisturbed.

Egg Collection: As soon as the female deposits her eggs, gather the newly laid king snake eggs to avoid dehydration, contamination, or damage. Rolling or rotating eggs can cause problems in their development, so handle them carefully to prevent this from happening.

Incubation Medium: To maintain consistent humidity levels and promote embryo development, place king snake eggs in a suitable incubation medium, such as hatchling substrate, vermiculite, or perlite. Achieve the ideal moisture content by weighing the incubation medium and adding water in a ratio of 1:1 or 1:2.

Temperature Regulation: To ensure optimal embryo growth and hatching, incubate king snake eggs at a consistent temperature between 78 and 84°F (25 and 29°C). A dependable egg incubator or a temperature-controlled space should be used to keep the temperature constant during the incubation process.

Humidity Control: To avoid dehydration and encourage optimal embryonic growth, keep the incubation chamber's relative humidity levels between 70 and 90 percent. To keep the ideal circumstances, periodically check the humidity levels and make any necessary adjustments to the ventilation or moisture.

Egg Turning: To avoid adherence to the incubation media and to encourage even heat distribution and gas exchange, turn king snake eggs gently and frequently throughout the incubation process. Once or twice a day, turn eggs, being careful not to agitate the orientation of the egg or disturb developing embryos.

Candling: Throughout the incubation process, regularly candle the eggs of king snakes using a bright, non-intrusive light source, like an LED bulb or flashlight. With candling, you can watch the growth of the embryo, find fertile eggs, and spot any anomalies in the embryo's development or signals of embryonic death.

Infant Care

The eggs of the king snake will hatch after 50–90 days of incubation, and the young snakes will come out of their shells. In order to guarantee the health, survivability, and environmental acclimatization of hatchling king snakes, neonatal care is necessary. Considerations for newborn care include the following:

Assistance with Hatching: Throughout the incubation stage, keep a close eye on the king snake eggs and be ready to help with hatching if needed. Using sterile scissors or tweezers, make a tiny incision or window in the eggshell to aid in hatching and avoid suffocating.

Housing for Hatchlings: As soon as possible after hatching, move young king snakes to a different cage that has the right humidity, temperature, and hiding places. To protect hatchlings from unintentional egress and to prevent accidental escapes, use compact, secure-lid cages.

Temperature gradient: In order to facilitate digestion and thermoregulation, keep a warm side of the hatchling enclosure at 80–85°F (27–29°C), and a cooler side at 75–80°F (24–27°C) for resting and chilling.

Humidity Regulation: To avoid dehydration and encourage healthy shedding, keep the hatchling enclosure's relative humidity levels between 60 and 70 percent. To keep hatchling king snakes at the ideal humidity levels, use a humidity box, misting, or moist substrates.

Feeding Schedule: After hatchling king snakes have shed their first skin and demonstrated that they are ready to eat, provide them with adequately sized prey items, such as pinky mice or newborn rats. To avoid regurgitation or choking, feed hatchlings every five to seven days with prey pieces little bigger than the widest part of their body.

Handling and Socialization: To prevent stress and encourage acclimatization to human contact, handle hatchling king snakes lightly and carefully. Provide brief handling lessons with steady, methodical movements to gradually foster confidence and trust.

Growth Monitoring: Weigh the hatchling king snakes, measure their length, and keep an eye on their feeding habits and frequency of shedding to track their growth and development over time. Maintain thorough notes on all feeding reactions, developmental milestones, and any anomalies or health problems.

Moral Aspects to Take into Account

Ethical concerns about genetic variety, conservation, and welfare are all part of responsible king snake breeding. Breeders ought to put their breeding stock's health and welfare first, follow recommended procedures for genetic management, and encourage conscientious ownership and care of king snakes raised in captivity. Some moral things to think about when breeding king snakes are as follows:

Breeding king snakes with an emphasis on genetic variety, species preservation, and conservation will help to preserve healthy captive populations while easing the strain on their wild counterparts. To promote long-term sustainability, take part in cooperative breeding projects, species-specific breeding programs, and conservation campaigns.

Welfare Considerations: Give breeding stock's health and well-being a priority. This includes providing them with suitable housing, food, veterinary care, and environmental enrichment. Steer clear of genetic modification, overbreeding, and inbreeding that could endanger the welfare or health of king snakes kept in captivity.

Transparency and Education: Give prospective purchasers clear information on the history, pedigree, and maintenance needs of king snakes raised in captivity. To encourage responsible stewardship and informed decision-making, educate customers about ethical breeding standards, responsible ownership practices, and conservation initiatives.

Genetic Management: To preserve genetic diversity and reduce the likelihood of genetic illnesses or defects in progeny, employ appropriate genetic management techniques, such as outcrossing, line breeding, and pedigree tracking. Steer clear of breeding people who have harmful features or recognized genetic health conditions.

Long-Term Planning: Develop breeding programs with long-term objectives in mind, keeping in mind the welfare, health, and genetic integrity of king snakes that will follow. Create breeding plans that put conservation of species, genetic diversity, and sustainable captive breeding methods first.

The intricate and satisfying process of breeding king snakes calls for meticulous planning, preparation, knowledge of reproductive biology, breeding behavior, and ethical issues. Breeders can contribute to the conservation, welfare, and genetic variety of captive king snake populations by adhering to best practices for species selection, environmental preparation, mating behavior observation, egg incubation, neonatal care, and ethical breeding standards. Breeders can make a significant contribution to protecting these amazing reptiles for the enjoyment of future generations by demonstrating a strong dedication to responsible care, skill, and hard work.

Chapter 7

FAQs for King Snake: Common Questions Addressed

Around the world, reptile aficionados are enthralled with the unique reptiles known as king snakes. Since they are well-liked pets, a lot of queries are frequently raised about their upkeep, temperament, health, and other facets of their lives. We'll answer some of the most common queries regarding king snakes in this extensive guide to help both new and seasoned keepers clarity and direction.

A king snake: what is it?

Native to North and Central America, king snakes are non-venomous constrictor snakes of the genus Lampropeltis. They are renowned for their eye-catching color patterns, which frequently include stripes or bands of opposing hues. As opportunistic hunters, king snakes mostly prey on other reptiles, such as lizards and snakes, though they also sometimes take in small animals and birds.

As pets, are king snakes good?

Yes, reptile lovers of all skill levels can have wonderful pets in king snakes. In general, they are temperamentally mild, hardy, and simple to care for. Kings may flourish in captivity and give their owners years of pleasure and company with the right care, handling, and enrichment.

What feeds the king snakes?

As opportunistic hunters, king snakes in the wild mostly prey on other reptiles, such as lizards, snakes, and amphibians. They may be fed a diet of small prey items, such as mice, rats, chicks, and even other reptiles, while they are in captivity. To maintain nutritional balance and avoid dietary deficits, a diversified diet is crucial.

When is the right time to feed my king snake?

Your king snake's age, size, metabolic rate, unique feeding reaction, and activity level all affect how frequently it eats. While adult king snakes should be fed once every 10–14 days, juvenile snakes may need to be fed more regularly, like once every 5–7 days. To keep your snake at a healthy weight, keep an eye on its physical condition and change the frequency of feedings as necessary.

How large can a king snake grow?

Depending on the species and subspecies, king snake sizes vary; some can grow up to 6 feet in length as adults, while others can only reach lengths of 3–4 feet. The majority of king snakes housed in captivity grow to be three to five feet long, while some may grow longer or shorter than these averages. It's critical to provide your snake enough room and environmental enrichment to match its size and activity level.

Are UVB lights necessary for king snakes?

Although king snakes do not technically require UVB illumination, giving them access to natural sunlight or additional UVB lighting can have positive effects on their health and general wellbeing. UVB light can support the immune system, calcium metabolism, and vitamin D synthesis, all of which are important for the health and vigor of captive king snakes. An artificial UVB light source that is on for ten to twelve hours a day can assist simulate the sunshine found in the environment.

How should the enclosure for my king snake be built up?

The health and welfare of your king snake depend on a well-configured enclosure. Give your snake enough room to move around, climb, and explore in a safe enclosure. Incorporate hiding places for your snake to feel secure, like caverns or branches. Aspen shavings or coconut husk bedding are good examples of a substrate that maintains moisture. You should also maintain proper temperature gradients and humidity levels.

Do king snakes enjoy being touched?

Many king snakes handle well under moderate, confident handling, and they may even love human interaction, but individual preferences may differ. It's critical to handle sessions with composure and respect, giving your snake time to get used to your touch. Steer clear of handling your snake when it's stressed, as right after feeding or right before shedding, and support its body at all times to avoid hurting or uncomfortable situations.

How do I determine whether my king snake is ill?

Changes in hunger, weight loss, lethargy, irregular feces, respiratory problems (such as wheezing or discharge), skin lesions, or behavioral abnormalities can all be indicators of disease or suffering in king snakes. If you have any worries about your snake's health, speak with a veterinarian who is knowledgeable in reptiles and keep a watchful eye out for any indications of disease or discomfort. Effective health issue management requires early discovery and treatment.

What is the lifespan of king snakes?

King snakes can survive in captivity for up to 15-20 years with the right maintenance and handling. Optimizing the lifespan and quality of life of your snake can be achieved by offering it an appropriate nutrition, frequent veterinarian treatment, an appropriate habitat, and enrichment activities. If you want your king snake to live a long, happy, and healthy life, you must dedicate yourself to its long-term care and guardianship.

Can I house several king snakes in one enclosure?

Although king snakes live alone most of the time, given the correct circumstances, some species may tolerate living together in a tank with appropriate animals. To reduce stress, hostility, and competition, it's critical to give enough room, hiding places, and environmental enrichment while housing many king snakes together.

Keep a careful eye on group dynamics and be ready to remove people from the group if disputes occur.

How can I determine the gender of my king snake?

It might be difficult to identify the sex of a king snake, particularly when it is young or small. Probing, a practice used by skilled reptile keepers or veterinarians, is the most dependable way to sex king snakes. Hemiepenal pockets, which are present in males but lacking in females, are probed for presence and depth using a thin probe put into the cloaca of the snake. As an alternative, some species might show slight variations in size, form, or color between males and females; nevertheless, these traits might not always be accurate markers of sex.

Are king snakes compatible with other reptiles in a home?

Even though king snakes are opportunistic predators that might feed on other reptiles in the wild, given the correct circumstances, they can occasionally be kept in tanks with appropriate housemates. To reduce the risk of predation, competition, or stress, it is crucial to select species that are similar in size, temperament, and environmental requirements when considering cohabitation with other reptiles. Always keep a tight eye on interactions and be ready to remove people from each other if necessary.

Are king snakes nocturnal?

In the wild, king snakes can experience "brumation," or periods of decreased metabolism and activity, throughout the winter. Brûmation is a state of dormancy brought on by seasonal variations in temperature and daylight duration, not genuine hibernation. If the environment is altered to replicate the seasonal variations found in nature, for example by adjusting temperature and photoperiod, captive king snakes may display brumation behavior. However, brumation should only be tried in carefully regulated settings under close observation and supervision, as it is not essential for the health and welfare of captive king snakes.

Are king snakes hostile animals?

Though individual temperaments can vary based on handling, heredity, and environmental stressors, king snakes are generally calm and non-aggressive around humans. Although king snakes are unlikely to bite unless provoked or handled incorrectly, they may display protective behaviors when frightened or startled, such as hissing, bluffing, or coiling. King snakes are tame and acclimated to human contact with the right socialization, handling, and surrounding enrichment.

Are king snakes noisy animals?

King snakes are not as noisy as some other species of reptiles; they are usually peaceful creatures. When scared or upset, they might, nevertheless, hiss or release strong breaths as a defensive mechanism. In order to alert prospective predators, king snakes can also make rustling or rattling noises by vibrating their tails against dry leaves or other substrate. Usually non-vocal, these noises are used more for intimidation or deterrence than for communication.

Can my king snake be bred?

It takes meticulous planning, preparation, knowledge of reproductive biology, breeding behavior, and ethical issues to successfully breed king snakes. Make sure the male and female snakes are in good health, have attained sexual maturity, and are compatible for breeding before attempting to breed your king snake. To encourage breeding success and responsible management, factors like species selection, habitat preparation, mating behavior observation, egg incubation, neonatal care, and ethical breeding methods should be carefully considered.

If my king snake stops eating, what should I do?

Anorexia, a transient decrease of appetite, is prevalent in king snakes and can be brought on by stress, environmental changes, temperature swings, or suppression of the reproductive season. See a veterinarian who understands reptiles if your king snake stops eating for a lengthy period of time or shows other symptoms of disease or discomfort, like weight loss, lethargy, or unusual behavior. This will help you identify the best course of action. To address the problem and improve your snake's appetite and general health, you may need to consider supportive care, veterinarian intervention, or changes to your snake's husbandry.

How do I deal with hair loss?

King snakes naturally go through a process called ecdysis, or shedding, in which they regularly lose their old skin to make room for growth and get rid of dirt or parasites. Your snake's skin may seem dull, opaque, or hazy during shedding because new skin is growing below. To encourage shedding and avoid dryness or retained shed, give your snake a moist hiding place and water the enclosure frequently. It is best not to handle your snake while it is shedding since it can have blurry vision and its skin might get irritated or sensitive. Give your snake privacy when it sheds, and if necessary, offer a shallow water dish for soaking to help soften the old skin and make removal easier.

What symptoms indicate that my king snake is gravid?

In female king snakes, gravidity, or pregnancy, is indicated by enlargement or swelling of the body, particularly in the posterior region close to the cloaca. Additionally, gravid females may display altered behavior, such as reduced activity, heightened appetite, or searching for potential nest locations. Keep a watchful eye out for any indications of pre-ovulatory swelling, follicular growth, or ovulation in your female king snake, as they suggest the imminent deposit of eggs. Provide an appropriate location for the nest to facilitate egg laying and reduce stress during the gravid stage, such as a humid hide box filled with moist substrate.

King snakes are intriguing reptiles that appeal to reptile aficionados as pets because of their distinct personalities, habits, and care needs. You can give king snakes the best possible care, enrichment, and stewardship by being aware of the frequently asked questions and worries about these amazing reptiles. Whether you are a first-time keeper or a seasoned breeder, maintaining the health, welfare, and happiness of your pet king snake requires knowledge, perseverance, and dedication. Your king snake can flourish and provide you with years of interest and delight as a beloved pet if given the right care and attention.

Chapter 8

In conclusion

We have learned about the fascinating world of king snakes in this extensive book, covering everything from their care, husbandry, and breeding issues to their natural history and behavior. As our adventure comes to an end, it is evident that king snakes are incredibly amazing reptiles that have captivated the minds and hearts of reptile aficionados all around the world. There is plenty to admire and enjoy about king snakes, regardless of your level of experience with these magnificent serpents—from newbie keepers just starting out to seasoned breeders committed to their welfare and conservation.

With their eye-catching color patterns, calm disposition, and wide variety of species and subspecies, king snakes provide a lot of options for reptile enthusiasts. Every species has its own distinct appeal and fascination, ranging from the vivid colors of the California kingsnake to the delicate elegance of the Florida kingsnake. We are fortunate to have the chance to see and engage with these amazing animals as keepers, learning from them and creating deep bonds that improve our lives.

It is our responsibility to provide king snakes the best possible care and stewardship to ensure their wellbeing, health, and quality of life. This duty goes beyond providing them with the necessities of food, water, and shelter and includes socializing, enrichment, and environments that mirror their natural habitat and advance their mental and physical health. We may establish habitats that support their growth and wellbeing in captivity by learning about their requirements, preferences, and habits.

Seeing the king snakes' natural interactions and behaviors is one of the most satisfying parts of owning them. A variety of actions exhibited by king snakes, such as hunting, foraging, basking, and exploring, provide information about their ecology and biology. We can better understand their place in their original ecosystems and the significance of conservation efforts to preserve their habitats by giving them the chance to exhibit similar behaviors in captivity.

It is a privilege and a duty to breed king snakes, and it calls for meticulous planning, preparation, and ethical thinking. We may support the welfare and general well-being of individual animals while also advancing the conservation and genetic variety of captive populations by taking part in ethical breeding projects. We can guarantee that next generations of king snakes are strong, healthy, and genetically diverse—ready to continue their heritage for many more years—through selective breeding, genetic management, and education.

We are reminded of the value of curiosity, compassion, and respect for all living things as we immerse ourselves in the world of king snakes. King snakes serve as a constant reminder of the wonder and diversity of the natural world as well as the interdependence of all living things, whether they are slithering through the underbrush in their natural habitats or curling up in the warmth of their enclosures in our homes. It is our duty as their caretakers to safeguard and conserve these amazing reptiles so that future generations can enjoy and appreciate them.

In summary, king snakes serve as more than just house pets or recreational animals—they are symbols of the great biodiversity of our world, ambassadors for their species, and sources of inspiration and awe for anyone who come into contact with them. We may develop a deeper understanding and appreciation for these remarkable reptiles and the environments they inhabit by embracing the majesty of king snakes and appreciating their role in the natural world. Our shared love and dedication to the survival and well-being of these amazing serpents unites us whether we are raising them in captivity, keeping them as pets, or trying to conserve them in the wild. May we approach king snakes with humility, gratitude, and reverence as we continue to learn and grow in our understanding of them, appreciating the honor of sharing our lives with these amazing animals.

www.ingramcontent.com/pod-product-compliance
Lightning Source LLC
Chambersburg PA
CBHW050111230526
45470CB00004B/1785